A SOMEWHAT FICTIONAL ACCOUNT OF A SEAFARING PIONEER

the Last American Whaler

DENNIS HUNTER

© Dennis Hunter. All rights reserved. No part of this publication may be reproduced, distributed, or transmitted in any form or by any means, including photocopying, recording, or other electronic or mechanical methods, without the prior written permission of the author, except in the case of brief quotations embodied in critical reviews and certain other noncommercial uses permitted by copyright law.

ISBN (Print): 978-1-09837-454-9
ISBN (eBook): 978-1-09837-455-6

Table of Contents

Preface 1

Prologue: Present Day 3

Chapter One: Central Oregon Coast • 1930-1931 5

Chapter Two: Skipper's First Boat 13

Chapter Three: Pacific Ocean, Off the Central Oregon Coast 19

Chapter Four: Waldport and Coos Bay, Oregon • 1936-1939 23

Chapter Five: Present Day 25

Chapter Six: Newport, Oregon • 1940-1944 27

Chapter Seven: Fields Landing, California • 1948 33

Chapter Eight: First Whaling Adventure 39

Chapter Nine: Good and Bad Fortunes • 1948-1951 43

Chapter Ten: Present Day 47

Chapter Eleven: Headed for South American Waters • 1952-1954 49

Chapter Twelve: Whaling Begins Again in
Northern California • 1958-1959 53

Chapter Thirteen: Young Sons Come Aboard 59

Chapter Fourteen: Present Day 63

Chapter Fifteen: Youngsters' Whaling Adventures Continue 65

Chapter Sixteen: Historic Flood and Whaling
Takes a Severe Blow • 1964-1971 73

Chapter Seventeen: Present Day 77

Crossing the Bar 79

Afterward 81

About the Author 83

*The sea,
this truth must be confessed,
has no generosity.*

—JOSEPH CONRAD

PREFACE

"Ever since the beginning of time, man has pitted himself against the power of the sea to learn its secrets and solve its mysteries. Many stories have been told of the ships and the men who sailed them; of sea beasts and the men who hunted them."
—from the movie trailer for the 1956 Movie: "Moby Dick"

Once there were as many as 750 whaling ships in the United States when the industry was at its peak. In those days whaling was a substantial part of the economy of such New England communities as Nantucket and provided the base for some of the great American stories of adventure on the high seas. Who could forget the Herman Melville tale, written in 1851, of Captain Ahab and his search for the giant white whale *Moby Dick*!

Commercial whaling in the United States dates back to the seventeenth century, but the industry in the twentieth century dwindled down to a precious few whaling adventurers on the northwest coast of California.

This is a story of one of those adventurers, Captain Gib Hunter, a true American whaling and fisheries pioneer of the twentieth century as seen through the eyes of his oldest son. It is a somewhat fictional account which attempts to stay true to events as remembered. But as memories are fading over the years, the author has attempted to write a story that will entertain as well as educate the reader.

The book contains vivid descriptions of whaling which by today's standards might seem barbaric, but please remember that until 1971, whaling was indeed an important part of our history. Each whale was killed humanely. In fact, strict regulations limited the taking of whales under a certain size and only 100 whales could be taken by each boat during a six-month season.

Up until the 1970's, whale meat was processed into pet and livestock food. In fact, for a time, whale meat was substituted for horse meat to produce dog food. Oil from the sperm whale had high demand and was utilized in cattle feed, industrial hydraulics and perfumes while the ground-up whalebone went into poultry food. No part of the whale would go wasted.

In 1971, the Federal Government called for an ending of American whaling. The Marine Mammal Protection Act "made it illegal for any person residing in the United States to kill, hunt, injure or harass all species of marine mammals, regardless of their population status."

Therefore, on December 31, 1971, when Gib Hunter stepped off the deck of the whaler *Dennis Gayle,* he became known as the last American whaler.

Prologue

PRESENT DAY

As the sun was starting to set on the vast Pacific Ocean, the old man was not happy. The listless sea was beginning to churn into something he could not control. He knew the ocean flowed with different personalities. Calm and cool one moment; furious and ferocious the next. In his seventy years at sea, he could begin to read the various currents and could predict what the ocean had in store. But, this was one day the mighty Pacific had a terrible surprise. She was about to lose her temper in a big way.

The waves were getting taller, rapidly growing to 10, 15, 20-feet and higher. From the deck of the small fishing boat, they looked as high as three-story buildings. Imagine slowly climbing all those stairs and then suddenly falling into nothingness until you slammed into the raging sea! The Pacific Ocean was definitely throwing a tantrum! Gale force winds started to toss the old man's 55-foot craft around. All the signs of a big storm were there. The ominous dark clouds began to form in the west, monstrous sculptures that portended impending doom. Sheets of lightning blazed in concert with the roaring thunder. As the boat was tossing and turning, the old veteran skipper tried to keep a weather eye out. He had just listened to the frequent weather reports on the radio which were opposite of the picture he saw of an angry sea. *Damn! What the devil were they thinking, putting out that latest report*, he thought. *I should have listened to my gut and headed in much earlier.* Everything he had learned and had been taught told him that this was going to be a much worse storm than expected. These rapidly-deteriorating conditions could close his path home and reminded him of a storm that happened on his first ocean adventure. As the boat's Captain, Gib Hunter felt a deep responsibility to get his crew, asleep in their bunks, safely home. He wondered if they could make it. There was always hope. But then, the mountains of water came.

Chapter One

CENTRAL OREGON COAST • 1930-1931

Hope. The old skipper's mind drifted back to when he was just twelve years old at which time he first felt the first pangs of yearning for the sea. Gib Hunter and his family lived on a houseboat along the Alsea River, several miles inland of Waldport, a small fishing port on the central Oregon coast. The houseboat wasn't much to look at, but it floated and was roomy enough for the family of five. It stood two stories high with three bedrooms, a small galley, bathroom and living room highlighted by the rustic Captain's chair and steering wheel as its most becoming features.

Gib, his brothers Bill and Glenn and their sister Agnes started their day at the breakfast table as they did every morning. By this time, their father, Will, had already headed for work. Their mother, Flora Mae, insisted on making sure that her kids began each day with home-made oatmeal with plenty of sugar and butter on top. Flora Mae was wearing her favorite pink and white apron, as always, when she served breakfast. She had a very pleasant matronly look, but her children always knew when she meant business! "Mom! It's too sunny outside to have oatmeal", said Gib. Looking toward the others, he started to stand up and said, "C'mon guys, let's get going!"

His younger brother Glenn, with a mouthful of oatmeal, glanced toward Agnes and said "Wait a minute. I don't want Agnes tagging along. She'll just slow us down. We don't need girls!" As the three boys started to leave the table, Agnes was a little put out! "Whatcha' mean? I wanna go!" she cried. Tears started streaming down her cheeks.

Flora Mae staved off any argument by persuading Agnes to stay behind. As she wiped the tears from Agnes's face, Flora Mae gently held both her cheeks, looked her in the eyes and whispered with a wink, "We gals have something special to do, don't we, Agnes? Let boys be boys! Know what I mean? We're going shopping in Newport. Then, I'll show you how to make a lemon meringue pie." Flora Mae was famous for her lemon meringue pies and she was pleased that she could pass along her recipe to her young daughter. Agnes immediately stopped crying and thought, she didn't want to go with those three hoodlums anyway! At that moment, the three young boys started to jump away from the table, Flora Mae stopped them and said "Sit down and hold your horses, boys! The day will wait. Finish your oatmeal. It's good for you!" How many times had they heard that line before? But, the boys knew never to say no to Flora. Reluctantly, they ate their oatmeal.

A few minutes later, after they dutifully kissed their mom goodbye and glared at their younger sister, the three brothers were skipping along the streets of the small town of Alsea, kicking cans back and forth along the sidewalk with Glenn yelling "Don't step on the cracks!"

Don't step on the cracks! You'll break our mother's back!" Neighbors, seeing the boys hopping and jumping down the street would shake their heads and smile.

Alsea is a small community of 600 people nestled in Oregon's scenic Coast Range on the banks of the Alsea River just upriver of Waldport, along the central Oregon coast. Timbered hills and pastured valleys surround the town, located on Highway 34 east of Waldport. Visitors to Alsea often comment on the abundance of natural beauty, including Alsea Falls, and the friendliness of its people. Alsea's most famous landmark is a beautiful covered bridge just two miles west of the town. It's called the Hayden Bridge. Sometimes, the three Hunter boys would fish off that bridge. There wasn't really much to the town which meant that everybody knew everybody. Like other small towns, this town had a Main Street which probably resembled thousands of early 1930's Main Streets across the country. On this brisk morning, the gray fog would weave its way through the small hilltops while the cold sea breeze coming up the river made the day start out cold and brisk. It rained a lot during the winter, but Gib felt lucky that the weather was good to him. The sun was starting to break through and it looked like Alsea was in for a great weather day. Little did Gib know that weather would play a crucial part in his life years later!

His father, Will, managed Waldport Seafoods, a small fish and crab processing plant over on the coast. After school and during the summers, Gib, along with his two brothers and sister, mostly helped their mother in her restaurant and pool hall. It was aptly named Hunter's Place and was located on a corner of Alsea's Main Street. In the restaurant, Flora Mae baked lemon meringue and banana cream pies that were famous along the central Oregon coast and attracted customers from miles around.

Hunter's Restaurant and Pool Hall in 1930's Alsea, Oregon

Sometimes, older brother Bill would drive the boys west on Highway 34 over the Hayden Bridge and on to Waldport to watch their father negotiate with the fishermen on prices. Gib could hear the loud arguments going on behind closed doors, but Will usually ended up setting fair prices and the fishermen happily headed to make their boats ready for sea. Gib was fascinated by the fishermen on the docks mending their nets and crab pots with miles and miles of rope and gear stretched out along the dock.

On that memorable brisk morning in Alsea, on full stomachs after breakfast, the three Hunter boys headed out to cruise the town on foot. After a few games of hopscotch and kick the can, boredom set in and they decided to see what was happening down at the pool hall. By this time, their mom and younger sister, Agnes, were already there. Cars were lined up in from of the place – cars they had never seen before. "I wonder what's up?" said Bill. When the boys walked in, they saw that the restaurant was full and there was a buzz of excitement in the air. What they didn't know was that a regional pool tournament final was going on that attracted people from the coast and inland to their parent's restaurant and pool hall. The boys looked around and noticed several strangers over at the two pool tables in the corner. "Geez!" said Gib. "Somebody's got my favorite table and I wanna play some pool!" He started to make his way over toward the tables when one of the players, a burly man with a sea captain's hat worn low on his brow, saw him coming. In Gib's 12-year-old mind, that man looked like one tough son of a bitch! As he approached, the older man slammed his pool cue down on the table and yelled "Get out of here, you unworthy brats! You have no damn business in here! You're distracting us! Now, scram!" One would think that these three young boys would turn tail and run, but no!

Young Gib stood his ground. His mind was made up. This was his parent's place and, by God, he was going to play some pool! "Give me a stick and I'll show you how pool is played", he said to the burly man.

His two brothers looked at each other in dismay! Bill came over and put his hand on Gib's shoulder and pleaded, "Gib. What are you doing? Let's get out of here. These guys are too big." Glenn shouted, "They'll kill us!" But, the scrawny 12-year-old Gib, was too stubborn and wouldn't back down. He continued to glare at the burly man and said "Give me a cue stick!" The burly man then swaggered over to young Gib, kneeled down and said through a liquor-filled breath: "Kid. You got guts! That's for sure." Grabbing the young boy by both collars, the man said: "Tell you what, son. You run the table and I'll show you my new fishing boat. Hell, I might even make you a deck hand. Rack 'em up!" The burly man smirked at his pool partners as if to say "Watch this! I'll show this little punk!"

The burly man laughed and pointed out that his young opponent was so small, he needed a box to stand tall enough to play, "Hey kid," he said. "Try standing on this box so you can see what the hell you're doing. I'll even let you break." The air in the pool parlor suddenly chilled because the strangers knew the burly man didn't like to lose. But young Gib, with a steady hand, knocked a few solids in and, as promised, ran the table.

The burly man was awed. What he didn't know was that Gib and his brothers Bill and Glenn played pool every day at this pool hall on this very table and, even at their young ages, played it like experts.

"You're a lucky son of a bitch, kid! Bet you can't do that again!" bellowed the big guy. Gib once more stepped to the table, stepped up on the box and racked again. He aimed the cue stick and gave it a mighty swing- as mighty as a twelve-year-old could make.

This time he knocked in some stripes and, once again, ran the table. The burly man was speechless. All he could do was stare at the scrappy kid had beaten him – twice! The strangers started to back away, thinking the burly man would tear the boy apart. Brothers Bill and Glenn were ready to run to get their mom. But the burly man came over and knelt down by the boy and said through bad breath and ragged teeth, "Well, what do you know, kid! I'm a man of my word. You're a hell of a pool player. Let's see what kind of fisherman you are. You get your ass on the dock in Waldport tomorrow morning at dawn. If you work as well as you play pool, we'll get along fine." Gib nodded in agreement, as he so wanted to be a part of the fisherman's world. He couldn't wait!

Sure enough, keeping his word, the very next day Gib headed for the dock. It was downright cold, but he dressed warm for it. Flora made sure of it. She again fed him full of oatmeal and sent him on his way. Fog was seeping along the Alsea River and the birds were still as young Gib made his way down the river in his small skiff past the Hayden Bridge to the docks at Waldport. Sunlight was just starting to appear through the fog behind him as he made his way to the dock.

After tying up his skiff, Gib climbed the ladder. He arrived precisely at dawn. Seeing him and breaking the eerie early morning silence, the burly man, whose name was Bud Burchit, yelled "Hey, kid! Git over here! Got a job for ya'!" Gib ran over to the boat and expected some type of glamourous job. He could imagine steering this boat past the rocks, through the jetties and out to sea. But he learned in a hurry that in the fish business you start at the bottom, not at the top so he spent the first days of his fishing career learning how to swab the decks of a fishing boat. Geez, he thought, this is backbreaking work! And, the smell of rotten fish! Yuck! However, Gib was smart enough to realize if swabbing the decks of a fish boat would lead to a life at sea, it would be worth it.

That night, it was all the youngster could do to row his skiff back up the Alsea Bay in the dark to the Hunter houseboat battling the current that was working against him as the tide was heading back out toward the sea.

Finally, he spotted the Hayden Bridge and he knew he was almost home. His back was aching and his legs were quivery after his first day on his new job. But, he made it and tied his skiff up to the port side of the houseboat. As he walked in the front door, Flora Mae exclaimed "Gib, you're walking like a stooped ole man! You got a sore back or something?" Gib, proud of his first day of swabbing decks, tried to make light of it, but his mom always seem to know best.

"Here, let's rub this liniment on your back. It'll make it feel better. And, I have a nice hot stew all ready for dinner" she said. She rubbed his back, gently, remembering how many times she had done the same for Will, her husband, during the early days of his fishing career. She then put dinner on the table, smiling, as the boys put away healthy portions of the warm, comforting stew. Then, right after dinner, Gib fell asleep on the couch. His father, who had made it back from the coast that day, smiled. Looking at his youngest son, he said "Well, that young whippersnapper sure worked hard today, didn't he, Flora?" Then, the proud father picked up his young son and carried him to bed. He watched him sleep for a few moments, turned off the lights and went to spend some time with Flora.

Chapter Two

SKIPPER'S FIRST BOAT

As the days and months went by, Gib started wondering why he was swabbing decks of someone else's boat when he should have a boat of his own. He confided to his brothers one night, "I'm going buy my own boat. Wanna be my crew?" His older brother, Bill, laughed, then said "Gib! You're so short, you can't even jump onto someone else's boat. How are you going to have one of your own?" Glenn, who was also taller than Gib, also started laughing at the thought of his older brother skippering his own boat. "Ridiculous!" he said. Agnes, over in the corner, just grinned. She knew that once her brother made his mind up to do something, he'd do it.

So, Gib set out on a course to buy his own boat. First, as he learned everything he could about fishing, he got ahold of a few crab pots that were thrown away but could easily be repaired. He learned that, with the proper bait, he could fish for crab from his small skiff. Thus, at the ripe old age of 14, Gib Hunter was pulling crab pots from the Alsea River.

In the late afternoons, one could drive along the Alsea Highway and see a short youngster on the side of the road with boxes of crab selling them to anyone who would buy them. Gib would yell "Here's some fresh crab, just caught early this morning, only a dollar a dozen!" Along the highway, people stopped, seeing this youngster with crabs for sale and would buy some. It didn't take very long before he started making some pretty good money.

Some days when Flora Mae wasn't busy working in the restaurant and tending the home front, and Gib wasn't catching or selling fresh crab, she would take him and his brothers and sister down to the coast. They would spend hours high above the Pacific Ocean on the cliffs near Sea Lion Caves looking out to sea where, as the sun went down over the horizon, they could see little lighted dots lining the horizon. "What're all those lights out there?" Agnes asked. Flora Mae would explain that these were crab boats picking up their pots before heading back to port. "Gee mom!" Gib said. "Looks like another highway down there, doesn't it?" Flora laughed and said, "You're right, Gib. It does!"

The days passed and one day, in Waldport, Gib happened to see a small derelict boat almost hidden away up on pallets in a dark corner of the dock. He couldn't wait to get home, excited to share with his dad what he had seen. "Dad. I got something to show ya", he said breathlessly. "I found the boat! I've been looking for one and I think this is it. It's in Waldport. It's old and needs a lot of work, but I think I'm gonna buy it." Will, knowing his son's mind was made up, grabbed his hat, and told Gib he'd look at it, but didn't expect much. "But, if it'll satisfy your curiosity, let's go see this old scow", he said! When they got to the dock, Gib proudly pointed out the boat. He looked expectantly at his dad. Will took one look at this derelict and gasped. Trying not to laugh, he said to his son "Are you kidding? Who's going to fix this up? I certainly don't have time! And, don't think for a minute that I'm buying a goddamn teenaged kid his own fishing boat!" Gib didn't hear his dad's sharp words. His mind was already made up. He had saved almost enough money selling crab to buy his first boat and his brothers said they would help him make the boat seaworthy.

Agnes said she'd help paint. But, Will was firm. When they arrived back home on their houseboat, Gib, heartbroken and head bowed, went into his bedroom without a word, thinking his dream of owning his own boat had been shattered.

Meanwhile, out in the living room, Will sat in his rocker and thought about his young son grieving in his bedroom. After conferring with Flora

Mae, he chuckled to himself. He had to admire the initiative shown by his young son. Finally, he yelled "Gib. Get out here! Now!" Gib came out of his bedroom and faced his dad, trying to hold back his tears. "Yes, dad", he blubbered. Will faced his son and said "Gib. I know you've been dreaming about this for a long time. I'm proud you want to continue in the fish business. Your mother and I talked about this and have made up our minds." Will then reached into his wallet and pulled out a one hundred dollar bill. As he handed it to Gib, he said "Here's the first $100 toward your boat. I know, with what you have already saved, you probably have enough to buy it and fix it up. But, don't ask me to help you. I don't just don't have the time." Knuckling the tears out of his eyes, Gib carefully took the $100 bill from his dad knowing how hard his dad worked for it. "Thanks, Dad", Gib said while wiping the tears from his eyes. "I know how hard you worked for it. I'll make you proud of me. I promise." Then, Gib headed for bed and dreamt of a life far out on the Pacific Ocean harvesting the bounties of the sea.

Will arranged to have the old boat put on a piece of land right next to Waldport Seafoods where it was easier for Gib and his brothers and sister to work on it. He could also keep his eyes on his children and make sure they didn't get into trouble. After school, the youngsters would head to Waldport and start working on the boat. They had to strip off all the old coat and sand all the hull boards before they could begin painting it.

One afternoon, their father, witnessing their efforts, said "Hey guys! This old boat has been out of the water for a long time. I brought you some caulking. You need to caulk all those boards clear around the hull. You don't want this old boat to sink!" The siblings looked at each other, wondering what in the world they had got themselves into.

It looked like Bill, Glen and Agnes were ready to quit, but Gib convinced them to stay. "Let's finish what we started", he said. "Let's make this the best boat in the fleet!" The other three youngsters gradually caught Gib's enthusiasm and the work continued.

Sometimes fate deals a winning hand or, in Gib's case, a winning pool match! Remember Bud Burchit, that burly man in the pool parlor? After Gib bought the derelict boat, the big man continued to mentor and encourage Gib in his endeavor and gave him some pointers on fixing it up. Bud and Gib would become lifetime friends.

One morning, as they worked on the boat, Bud Burchit asked "Hey Gib. Whatcha' goin' to name her?" Gib thought back to those hazy days of watching distant crab boats off the coast looking like dots and said "I'm naming her the *Dot*." So, every weekend, Gib and his two brothers and sister with Bud, would head down the Alsea River in their little skiff to Waldport and worked feverishly to make the *Dot* ready for sea.

One day, while the boys and Agnes were working on caulking and painting the hull, Bud drove up in a big pickup towing a trailer with a huge load covered with a tarp. "Surprise!" the burly man yelled. "Look what I brought!" Gib came over to the truck and wondered out loud, "What's under that tarp, Bud?" Bud said, "A diesel engine which would be perfect for this boat. I'll even install it for you, Gib." "Whoa! Wait a minute, Bud", Gib stammered. "I can't afford a new engine." Bud laughed and said "Gib, remember my buddies that watched you beat me at pool? We played poker one night and I won a brand new engine. I'm installing it on my boat and was trying to figure out what to do with this engine that I took out of my boat when it suddenly dawned on me: it would fit perfectly on this boat. The engine's in great shape, so, here it is!" Gib was flabbergasted but accepted the gift with sincere gratitude.

While Bill, Glen and Agnes continued their work on the hull above, Bud and Gib headed down to the engine room to install the big diesel. Gib learned a lot that day thinking "Boy! This is sure a lot of work, but it's worth it!" He could picture in his mind his new life on the sea.

A whole year went by and finally the *Dot* was fully equipped and ready to go! The big day to launch her was here and the whole family traveled down the river to the docks of Waldport to watch. The day started with a thin mist, but it burned off quickly to a clear blue sky. It was another good weather day for the young skipper. Flora and Will stood by the launch ramp watching and worrying. As the boat slowly creeped down the ramp and into the water, everybody had their fingers crossed hoping that it wouldn't sink. Gib kept worrying that all that caulking wouldn't hold and the boat would just continue to slide beneath the bay and that would be the end of his dream. But, the boat smoothly glided into the bay without any hitches, with Gib proudly standing on the bow waving to his family on shore. The *Dot* looked great. All the family's hard work paid off. Her fresh paint glistened in the bright sun. Her crab pots were neatly stacked on the back deck. Her like-new diesel engine purred like a kitten. She was a 40-foot trawler ready for sea!

As the *Dot* was tied up at the dock and the successful day ebbed like the evening tide, Will smiled. If the buttons on his vest could burst, they most certainly would have as he was very proud of his son's accomplishments. But, he whispered to Flora Mae with a twinkle in his eye, "I damn sure will never admit it. Not in front of him, anyway."

Weeks later, it was time to take the *Dot* on her first fishing voyage. Bud, the burly man, and his pool partners, who were veterans of the sea, joined the inexperienced young boy for the boat's maiden voyage. "Couldn't have a 14-year-old skipper go out to sea without help, now could we"? mused Bud with a twinkle in his eye. Seeing Gib's siblings, Agnes, Glenn and Bill onshore, Bud commanded "Hey, guys! Make yourselves useful! Untie those lines will ya'? As Glenn and Bill started to loosen the lines, the skipper, Gib, yelled: "Wait a minute! I can't see a thing!" "Whatcha mean?" Glenn asked. Gib said "I can't see over this thing!" He was referring to the wind deflector on the bridge. Then, Will stepped to the rescue! Laughing, he ran to get a big crab box lying on the dock. He threw it on the deck of the boat and told Gib "Hey! Grab that damn box and stand on it, you little whippersnapper!" Gib did as he was told and his long career on the high seas began while standing on a crab box. Will and Flora, with Agnes, Glenn and Bill by their side, watched as their young skipper made his first trip to sea.

It was an early Sunday morning when the *Dot* weaved its way slowly passed the jetty and out into the Pacific Ocean. Gib, Bud, and crew members, Johnathan Moore and Alex Johnson, looked forward to the short trip. The crab pots were stacked and the search began for the elusive Dungeness crab. The weather was calm with flat seas so the crew didn't pay attention to the warning signs of the weather forecast. Too bad! It was about to get ugly!

Chapter Three

PACIFIC OCEAN,
OFF THE CENTRAL OREGON COAST

As the *Dot* made its way out to sea, the crew ate a big breakfast of pancakes, eggs and sausage with a steaming cup of coffee while Bud was up in the wheelhouse heading the boat to what he hoped would be fertile crab grounds. "Hey Gib! Take this plate of pancakes and cup of coffee up to the wheelhouse for Bud, will ya'?" asked Jonathan. Gib, impatient because he didn't want his pancakes to get cold, said "You take 'em, Jonathan." The veteran crewman scowled at the youngster and that's all it took! Gib carefully made his way up the stairs with both hands full and delivered breakfast to the real Captain. He was lucky he didn't spill anything with the way the swells were building.

Bud decided that the crab grounds were right along the 40-fathom line on the edge of the Continental shelf and were probably abundant with Dungeness crab. He picked out a spot about three miles off the breaker line off Heceta Head, about 25 miles south, so that's where he was headed to dump their pots. In the wheelhouse eating his breakfast, Bud looked at the coastline off to port, harking back to those seagoers who, over 160 years earlier, had navigated these treacherous currents to explore the rugged Pacific Northwest.

All of a sudden, Jonathan came running into the wheelhouse and broke Buddy's reverie. "Gib! Bud! Come here! Look off our starboard bow!" Jonathan yelled. They all rushed out to the bow and started to scan the ocean. Just a few yards away, a huge 35-foot humpback whale was blowing! "Well, thar she blows!" Bud said with a laugh. Then, off to the whale's left, another whale breached and sent a cloud of water skyward! Gib would never forget that sight and, of course, didn't realize how the humpback whale would be a part of his life years later.

It was in the afternoon, after half of the pots were unloaded, when the weather changed for the worst. Gib had a hard time concentrating because he saw the swells begin to build, and in the small wheelhouse, it was all he could do to hang on! On the back deck, crewman Jonathan was starting to get seasick, hanging his head over the stern.

Alex, used to rougher weather, laughed and said "what's the matter, Jonathan? You look pretty green. You getting ready to feed the fish? Too many pancakes, eh?" Up in the wheelhouse, Bud made a decision. He told young Captain Gib "I don't like the look of this. See those swells starting to build and look at those black clouds forming over there. That's a squall line. We'd better get out of here and start heading for the barn. Let's leave our pots soaking right where they are. We can always come back and dump the rest of 'em!"

By then, the swells started to batter the small boat. Waves began to smash against the small bow, water washing onto the decks. "Better get your survival suits on, men, just in case!" Bud shouted.

As Bud started to turn the wheel to head back to Waldport some three hours north, he saw smoke coming from the engine room and the engine went silent. "Shit! We have a problem. We're shut down!" Bud said. "Goddamn engine quit!" So, the crew of the *Dot* faced the mercy of the treacherous seas without any power to help them.

The currents started to turn the drifting boat into the trough, which means that instead of pitching from bow to stern, the small boat was rolling side to side. It wouldn't take much to capsize it. The crew would have to restore power soon or they would be thrown overboard into the frigid Pacific waters. The winds began to blow harder and the swells started to increase. Bud reached for the radio and called for help. "Coast Guard! Coast Guard! This is the fishing vessel *Dot*. We're west of Heceta Head just off the breaker line and we've just lost power. There's smoke coming out of the engine room, but I think we can put the fire out. We're dead in the water and might need a tow. But hurry before we smash into the rocks!"

The Coast Guard responded quickly. "We have a situation up at Yaquina. We'll get to you as soon as we can. It might be a few hours." Bud then sprang into action! He yelled to the crew "Hey! Get up here. We have a problem!" Jonathan and Alex made their way to the wheelhouse, almost crawling because the rolling was so bad. "Look guys! Unless we can get out of this trough, we might roll over. Alex, take the life rafts and get ready to roll 'em over the side. We might have to abandon this ole' girl!" exclaimed Bud. And, in the same breath, he yelled "Jonathan, take the extinguisher down to the engine room and put that damn fire out!" Young Gib, caught up in all this drama, started to sob. His dream of owning and captaining his own boat might be shattered!

Just then, as thunderclaps could be heard in the distance, a voice from the radio crackled, "Fishing boat *Dot*. This is fishing vessel *Karen L* calling. We're headed for Waldport and can tow you to port. We're only five miles away. Hang on! I should have you in sight in just a few minutes!" Waves of relief hit the members of the crew. They might be saved yet!

A few minutes later, Jonathan appeared in the wheelhouse with grease all over his face which also had a big grin. "I got the fire out, but the engine's going to need a lot of work", he said. Then, Alex yelled that he sighted the *Karen L* approaching from the southwest. Bud grabbed the radio, keyed the microphone and said "*Karen L*. We have you in sight. Our fire's out but we sure could use a tow. Come on in on our starboard side. It's more protected from the wind." As he said that, a big swell hit the boat and sprayed water all through the wheelhouse. "Grab me a towel, Gib!" said Bud. "Jonathan! Alex! Get on deck on get ready to throw lines!"

It took over an hour, but finally the two boats were hooked up and started their slow trek north to Waldport. Ironically, the *Karen L* was owned and operated by the Hall brothers: Ray, Wilburn and Chester. These same brothers would remain some of Gib's closest friends for over a decade.

Together, the Hall brothers on the *Karen L* and veteran skipper Bud Burchit on the *Dot* maneuvered the two fishing boats against a tough northwesterly swell and head winds of 25 knots. Five hours later, they made their way into the sheltered Waldport docks. The *Dot* and its young skipper, both on their maiden voyages, made it home safely after all.

Chapter Four

WALDPORT AND COOS BAY, OREGON • 1936-1939

As the years went by, up the river in Alsea, Gib and his brothers attended Alsea public schools. In fact, at Alsea High School, all three tried out for the varsity basketball team. Gib's two taller brothers made the team, but at 5 feet, 10 inches, Gib was too short so he went out for the school's baseball team, becoming the starting shortstop with a rifle arm. He played during his senior year and earned his letters and school jacket. His older brother Bill starred on the varsity basketball team became an all-star and also earned his letters and school jacket. Glenn, a year behind Gib, played some football and also earned his letter. After his sports career ended, Gib's mind was on saving money and buying another boat. His mother Flora Mae made sure some of his mind was on education. "You'll never amount to anything without a high school education, Gib. I know you want to go fishing, but you need to graduate." Gib took his mother's advice, studied hard and, on a somewhat sunny late spring day, he finally graduated from high school. He actually ended up fifth in his class.

His dad and mom, Will and Flora Mae, made sure they had front row seats for the graduation ceremony. "No one's going to stop me from seeing my son graduate!" said Flora Mae. Agnes, Bill and Glenn were there, too. Even Will's brother Harold and his wife Della came from nearby Newport.

The boys referred to their Uncle Harold as Uncle Hunk. That name stood for the rest of Harold's life. He was a fisherman, too, operating a small boat out of Yaquina Bay. After the graduation ceremonies concluded, Will and Flora Mae invited everyone to Hunter's Place for a graduation party. But, Gib's mind was elsewhere – far out to sea! As spring turned into summer, Gib was ready to continue his adventures at sea.

In the meantime, the depth of the Great Depression hit the country. The commercial fishing industry was scrambling trying to sell fresh seafood wherever it could. Young Gib saw an opportunity here, but he would need a bigger boat.

One day while wandering the docks in nearby Waldport, Gib saw his old friend Bud. The burly man was excited to see his young protégé. "Gib! I hear you're looking for a bigger boat." said Bud. Gib, of course, responded "Yeah, I am!" Bud said "I heard that a new boat was being built down south in Florence. "I got nothin' going, so let's go take a ride down there and take a look at it". So Gib and his mentor, Bud Burchit, jumped into Bud's dusty pickup and headed for the boat building facility where they met a young Swede named Yonke Elving. Yonke and his friend Fritz Yost were building a new boat. The boat was 55 feet in length and had a rounded stern. It was designed for both crabbing and drag fishing and was a perfect fit for what Gib had in mind. "Whacha' doin' with the boat?" Gib asked. Yonke responded with a heavy Swedish accent "Ve're goin' to launch it and sell it. Should make one fine fishing boat for somebody. We named it the *Vinga*" Yonke added. The name was appropriate for young Yonke because the name *Vinga* actually was the name of a small island outside Gothenburg's harbor entrance in Sweden. The 19th century Vinga Lighthouse was a beacon in the waterway of the Swedish west coast and saved many lives over the years. It became a beacon of hope.

Gib made the deal right there on the spot with money he had saved selling crab and fishing. He also had a buyer for his first boat, the *Dot*. He told Yonke he'd name his new boat the *Winga*. This boat would remain with him the rest of his life.

Chapter Five

PRESENT DAY

As the rogue wave approached his small boat, the old skipper knew he had to do something fast! He swung the boat around to meet the challenging waters about to crash head on into the bow. Gib yelled downstairs to his crew. "Hey guys! Better get your survival suits on and get up here…now!"

As he concentrated on navigating through the rough water, Gib heard the alarm sound. He then looked at the gauges and saw the oil pressure go into the red. He knew his engine was about to seize up. He desperately needed all the power he could get to make it through this storm.

He reached for the radio. "Mayday! Mayday! This is the fishing boat *Winga*! We're in one hell of a storm and our engine is seizing up. We need help."

He yelled again for his crew. But, they never showed.

Chapter Six

NEWPORT, OREGON • 1940-1944

Over the next few years, Captain Gib Hunter really caught on to the fishing business. In fact, he was considered one of the "up and coming" young fishermen in the fleet. He moved the *Winga*, north to Yaquina Bay at Newport, Oregon. Along the harbor which stretches for a half-mile, you could hear the sounds of sea lions and smell the fresh fish and Dungeness crab being delivered to the processing plants. Some people would come in droves to buy fish directly from the boats. Framed in the background is the picturesque Yaquina Bay Bridge which marks the gateway to the Newport Bar, one of the most treacherous entrances to the Pacific Ocean on the west coast. The 600-foot semi-arch bridge was opened to traffic in 1936. At each end was built a pedestrian plaza, which provided excellent viewpoints of both the Pacific Ocean and the quaint Port of Newport harbor. Gib took this all in and could easily see that this was a great place to grow his fish business. He found moorage space on the harbor right next to the *Karen L* owned by the Hall brothers who helped tow Gib's first boat, the *Dot* to port years earlier.

Gib quickly realized that he needed to modernize his boat with up-to-date equipment so he called his friend Yonke and asked him to accompany him to Portland. He didn't know it at the time, but this trip would change his life. While waiting for Portland's big marine store to open, the two young men noticed a swimming suit contest going on down town. Gib said to Yonke "Hey! Let's go check it out!" Yonke responded with a grin "Yeah! But don't tell my wife!" While glancing at the models, Gib happened to notice a gorgeous brunette modeling Jansen swimming suits. In fact, she won the contest. "Yonke. Look at her! She's beautiful!" Yonke said with a chuckle, "her name is Maxine! She happens to be a first cousin of my wife, Ellen. I'll introduce you!"

Yonke called his Ellen and asked her to come up to Portland and help introduce Maxine to his friend. That night, the four of them had dinner in a nice Portland restaurant. Gib and Maxine hit it off that very night and were married in the summer of 1941 and the rest is history.

Maxine and Gib Hunter on their Wedding Day – June 28, 1941

Gib and Maxine's wedding was spectacular! It was held on a rare beautiful sunny day near the Yaquina Lighthouse overlooking the beautiful Pacific. Most of the fishing fleet was there, including Bud, Jonathan, Alex and the three Hall brothers. Yonke and Ellen were best man and matron of honor. Maxine was dressed in an elegant simple cut suit and fisherman Gib looked dashing in a new tweed suit. The reception followed and then Gib and Maxine departed in their new black De Soto.

After their honeymoon, Gib and Maxine bought a small house with a beautiful fenced back yard located in the suburbs of Newport. They would often get together with their new neighbors, Yonke and Ellen. During that summer, Gib's parents, Will and Flora Mae, sold their Waldport and Alsea businesses and moved to Newport where Will worked for one of the processing plants on Newport's charming little harbor. Flora Mae continued to make her famous pies as she looked out the window at the beautiful Yaquina Bay Bridge.

Maxine always had the "boat" radio on in her living room and could tell when the *Winga* was about to enter Yaquina Bay. She would bundle up and head for the view point near the Yaquina Head Lighthouse and watch the boat come in. This 162-foot lighthouse stood taller than any on the entire Oregon coast. After she saw the *Winga* steam under the bridge, Maxine headed home in time to prepare a hot meal for her skipper.

One October day in 1941, while Gib and his crew were bringing a boatload of crab into Yaquina Bay, he felt like something sharp stabbed him in the stomach. He could hardly breathe. He yelled to his crew for help! Jonathan quickly came up to the wheelhouse and found Gib lying on the floor. Jonathan grabbed the radio and pleaded for help. The Coast Guard responded quickly and said they'd have an ambulance waiting at the dock.

The ambulance quickly delivered Gib to the hospital in Newport where the doctor correctly diagnosed a ruptured appendix. After the successful surgery, the doctor said "Gib. Had your crew not reacted the way they did, you would have died! They saved your life."

The doctor pleaded with Gib, with Maxine and Flora Mae looking on. "You need rest. I don't want you doing much for a couple of months" he said.

Two months later, on the morning of December 7th, Gib, Maxine, Will and Flora Mae were hovered around the radio listening to the broadcast of Franklin D. Roosevelt telling the nation that they were at war! Gib so badly wanted to enlist, but his draft board turned him down because of the hole in his abdomen caused by his appendix surgery. He was out of the war, but he and his fellow Oregon commercial fishermen helped provide much needed food for those fighting overseas.

Six months later, Gib and Maxine's family grew as they brought a new son into the world. Two years later, a little baby girl joined the family. Yonke and Ellen Elving also had a little girl, followed by a boy. The families would always get together to celebrate their children's birthdays.

Gib's mind was always on the sea! He had a responsibility to care for his family and would brave the rough bar of Yaquina Bay and head out into the Pacific. He never forgot to check the weather and the tides as the Yaquina bar was one of the most treacherous on the Pacific coast. He didn't want to get caught out at sea with the Yaquina Bar closing behind him so he made checking the weather a part of his routine. Maxine would continue her routine, too. She would bundle up her two children and head for the view point to watch her husband and his crew bring another catch of fish into the harbor. She would always worry because the little *Winga* looked so small out there in the huge Pacific. At the same time, Gib's brothers, Bill and Glenn formed a partnership with their father Will and bought a 60-foot boat and named it the *Pearl Harbor*. Sometimes, the *Pearl Harbor* would team up with the *Winga* to find the most bounty.

Some days, when the ocean was just too rough, Gib would spend time maintaining his boat and his equipment. One of these in-port days, four years later, when Gib was on the back deck of the *Winga*, a man about his age approached. He resembled the young actor Lee J. Cobb. His name was Herman Foland. He had a small boat docked in the same marina.

"Gib. I'd like to talk to you about a partnership. I've heard a great deal about you. You're one hell of a fisherman. There's a hell of a lot of fish to be caught down in Northern California. I'm thinking of moving to Eureka. It's a great opportunity and I'd like you to come with me."

Foland offered to host Gib and Maxine for dinner to talk over their plans. He introduced them to his wife, Helen. At dinner, Herman pointed out the abundance of fish and crab just offshore of Humboldt Bay, just ninety miles south of the Oregon-California border. "This could be a money maker for us, Gib", said Herman. Gib, never one to pass up an opportunity, pondered a bit and said "Let me talk to my wife, Maxine."

After much discussion, Gib and Maxine, with their two young children now aged three and one, decided to uproot and move 300 miles south to Eureka, California. The morning of the move, young Maxine was apprehensive. She had never been anywhere but Portland and Newport in Oregon. What was Eureka like? Where would the family live? Could Maxine abandon her mother and grandmother who lived in Portland? But, the decision was made. The family was bundled in Gib's black De Soto while his older brother Bill and his wife Clara loaded up their truck and trailer with Gib's family possessions. The trip south took a miserable thirteen hours, mostly around treacherous hair-pin turns and fog that covered the mountain peaks. At one point, between Gold Beach and Brookings, Oregon, Bill couldn't make it around a sharp turn and the trailer ended up in a ditch. Maxine watched as her family possessions spewed out of the trailer and almost over the cliff into the sea. After several hours, Gib and Bill finally got the re-loaded trailer back up onto the highway and the modern day wagon train continued its journey south without any further mishaps. Maxine, waiting in the De Soto, was sadly wondering if they had made the right decision. Tears coming down her face were not tears of joy. She left the life she knew back in Newport and the rest of her family lived in Portland. But she was determined to support her husband and his dreams that sent them south along the perilous roads of Highway 101.

After what seemed like an eternity, they finally reached the small town of Arcata and had their first glimpse of Humboldt Bay. Their destination – Eureka – was only an easy eight miles away. Herman had made reservations for the families at a motel. A restaurant was located right next to the motel where the travelers ate a hot meal and looked forward to a good night's sleep.

The city of Eureka, with slightly over twenty thousand people, is situated on the shoreline of beautiful Humboldt Bay. To the east, you can see the majestic coastal mountains with giant redwoods. The city is driven by rich fishing and lumber industries so the future of Gib, Maxine and family looked bright. However, their first few years in Eureka, they didn't have much money. Most of it was invested in the *Winga*. Maxine was still pining for the life she knew up in Oregon. But, Herman and Helen Foland convinced them to stay and helped them find a small place in the western part of town. The young family settled into its new life together and made the most of it. The first few years, Gib did fairly well fishing with the *Winga*, but several years later, his life took a gratuitous turn.

Chapter Seven

FIELDS LANDING, CALIFORNIA • 1948

Five miles south of Eureka lies Fields Landing. This small town had a lumber drying and shipping facility, a railroad and several fishing operations. From 1940 through 1952, it housed the only whaling station anywhere in the United States. One could always tell its location by the stench. "My God! It stinks" cried most Fields Landing residents. They were referring to a small whaling station at the south end of town. In 1948, the whaling station had about run its course as its owners were running out of money and were feeling the pressure of the townsfolk to do something about that smell!

One day, Gib and Herman walked into a cocktail lounge in Fields Landing and happened to see one of the whaling station owners, a guy named Max Hoffman, sitting alone at a table over in a dark corner. Herman, always an opportunist, whispered to Gib "Hey! See that guy over there in the corner. He owns the whaling station here in town. We could probably buy it for a song." Gib laughed. "You're kidding! Who'd run it? And, who would bring in the whales? Not us. We're fishermen" Gib exclaimed. Herman said "let's walk over to him and see what happens."

The two fishermen walked over to the table where Hoffman seemed deep in thought. Herman said to him "We might be interested in buying your whaling station.

Are you interested in selling?" Max was ready to sell, Herman was enthusiastic, but Gib wanted to think about it overnight.

Later that night, Herman confided to Gib "I've heard that the Navy has mothballed some boats after the war. There's a couple that'd make great whalers and I think we could get 'em cheap! They're only seven years old. We could hire a guy to run the station and we could go whaling. There's good money in it and there's a lot of whales right here off the coast. Finbacks. Humpbacks. Sperms. Maybe some Blues."

Wow! Gib remembered seeing these huge leviathans over the years, but never thought he would be hunting them. After much consideration, he thought he could hire a crew to run the *Winga* to keep it fishing and, if he could get a couple of whale boats, he might be rich! "Herman. Let's do it" said Gib.

On a brisk November morning, Gib and Herman made their eight-hour journey along curvy Highway 101 to Vallejo on Suisan Bay in the San Francisco Bay area. Along the way, they talked about their upcoming adventures hunting whales off the Northern California coast. When they reached the Mare Island Naval shipyard in Vallejo, they immediately saw the boats they were looking for – right in the middle of the mothball fleet. Gib ended up buying three Navy surplus inter-island cargo carriers.

The APc-5 surplus Naval Inter-Island Cargo Vessel
later to be named the **"Dennis Gayle"**

Each boat was equipped with a huge Superior 6-cylinder 400 Horsepower diesel engine which ran the boats at a slow 10 knots. Much later, in the 1970's, these engines were replaced by more economical diesels that weighed 4,300 pounds less and generated 600 Horsepower of power. Meanwhile, Herman found a surplus Navy minesweeper.

So, the two new whaling Captains bought the Fields Landing whaling station, the only one on the American continent. They hired an experienced manager to run it. The new manager, Newt Stewart, set up some bleachers, a display of whaling exhibits and more parking spaces to attract tourists to see the whales

Gib named one of his boats the *Dennis Gayle*, after his two children and Herman named his minesweeper the *Lynnann*, taking the middle names of his two girls.

Whalers **Dennis Gayle** *and* **Lynnann** *in Eureka in the early 1950's*

Another one of Gib's boats was named the *Allen Cody* was named after a famous Civil War veteran. The third AC-5 was sold to Gib's Uncle Harold who named it after his daughter *Donna Mae*.

On the bow of each diesel-powered boat, Gib and his crew had mounted Norwegian muzzle-loading harpoon guns that fired six-foot 180-pound harpoons about 100 yards. The tip, or warhead of the harpoon, contained a black powder charge with a time fuse designed to go off inside the whale. Near each harpoon's head, metal prongs were tied together. When the harpoon entered the whale, these prongs would break the thin nylon ties and spring open, sinking a firm hold on the whale. Each harpoon is connected with 4,000 feet of heavy nylon line and steel cable which would be wound around a winch for pulling whales closer to the boat.

The harpoon gunner, a short thin-haired man named Bill Bartow, had some experience in catching whales. Herman's boat, the *Lynnann* had a similar crew on board and was equipped similarly to the *Dennis Gayle*. The third boat, *Allen Cody* was equipped for drag fishing and temporarily joined the *Winga* and others in the Humboldt Bay fishing fleet. The *Allen Cody* would enter the whaling fleet later.

Gib Hunter and his gunner loading a harpoon on Humboldt Bay in the 1950's

Chapter Eight

FIRST WHALING ADVENTURE

On a spring day in 1948, the two whalers headed out of Humboldt Bay looking for their first catch of humpback, finback and sperm whales. Passing through the jetty on the Humboldt Bar, the two boats looked majestic - armed with their harpoons. They headed out to sea side by side and made quite a sight for the dozens of onlookers from each of the north and south jetties. Finally, they reached the sea buoy, about a mile outside the Humboldt Bar. Gib got on the radio: "Calling the *Lynnann*. Calling the *Lynnann*. This is the *Dennis Gayle* calling." Herman responded "This is the *Lynnann*. Over." "Ah, yeah, Herman" Gib said. "I'm headed south off Cape Mendocino. Heard there were a bunch of humpbacks down there. I'll let you know what I find. Over." Herman responded "I'm headed due west. Heard there were some big finbacks out there about twelve miles out in the blue water. Over." "Good luck and good hunting. *Dennis Gayle* out" said Gib. "*Lynnan*n clear" responded Herman. So, each boat steamed toward hope for new adventures, never knowing what might lie ahead.

It took the *Dennis Gayle* a few hours to reach the Cape. Because of the changing currents, this area was an abundant feeding ground for whales. Fortunately, the weather was calm and the skies were clear which made whale spouts, or blows, easier to see. The youngest crewmember headed up to the crow's nest to start looking. The job of being perched in the crow's nest looking out over empty horizons brought hours, or sometimes days, of boredom. But, fortunately, an hour into the search, a cry came from above. "Gib! Off our starboard bow, I see one!" So, the first whale spotted in Gib's whaling career was not announced by the infamous "Thar' She Blows", but instead: "I see one!"

Cranking the wheel to starboard, Gib increased speed and headed for the whale which was still blowing steam high into the blue skies. Bartow, the harpooner, raced to the bow to get everything ready for the shoot. Hunting whales was truly an art. It wasn't easy, but took a great deal of skill. The object was to maneuver a 100-foot boat to within 100 feet of the whale off the bow within about a 45 degree arc. Some whales were predictable. Some were not. But Bartow was ready. He stood on the bow, dressed for the cold weather, with a cigarette hanging out of one side of his mouth.

As soon as the *Dennis Gayle* arrived at the correct spot, the whale's fluke, or tail, shot straight up in the air, as the humpback dove for the bottom. Gib shut down the engine and waited. Some whales, specifically sperm whales, could stay down for over an hour. Usually, humpbacks stayed down around 10 to 12 minutes. Gib had the stop watch going. He started it when he saw the whale diving. The scene was quiet. Only the sounds of the boat meeting the Pacific swells could be heard. Every crew member had his eyes peeled. And then, they could hear a "poof". They turned their heads and saw the whale blow just off their stern. Gib cranked the wheel, added full power and yelled to his gunner: "Bartow! Damn thing's behind us!" The cat and mouse race had begun – man against beast!

Crewmembers had to learn to hold on as the boat made its drastic 180-degree turn in ocean swells. As the boat completed its turn, the whale dove again and Gib started his stop watch. The wait began again as the skipper maneuvered the boat anticipating where the whale would come to the surface. Ten minutes went by. Nothing. Eleven minutes. Nothing. Finally, on the twelfth minute, Bartow could see the whale coming up right in front of the boat. "Gib!" he screamed. "The whale's a couple hundred feet in front of us." Gib shut down power and the boat drifted to a stop.

"Dennis Gayle" *gunner lining up his shot!*

There it was. Surfacing right in front of them. What a sight! A huge 35-foot humpback. Bartow lined up his shot and……BAM! The harpoon shot out and struck!

The rest of the crew couldn't see because of the smoke from the cannon, but as the smoke cleared, they found they had a mad whale on the other end of the line. Now the chase was on!

After an hour of towing the boat offshore, the whale finally gave up. Thousands of feet of cable wound its way back onto the boat through a series of blocks attached to railroad springs to absorb some of the jerks of a struggling whale and to keep the strain steady despite heavy swells. Now, the process began to bring the whale back to Fields Landing. First, it was crucial that the whale remain afloat. A whale weighs about a ton a foot and, when killed, it would roll over and start to sink. Gib shouted from the bridge "Hey guys! Take up the slack with the winch so we don't lose him." When the dead 40-ton whale again reached the surface, a hollow pipe was jabbed into the body cavity and air was pumped into it to keep it afloat.

After it was filled with air, the whale was then tied fluke-first to some cleats located along the starboard side. With all that accomplished, it was time to head for home eight hours away.

After the hard work was done, Gib headed to the wheel house and set the course for Humboldt Bay. He picked up the radio to let Herman know he had a good-sized humpback alongside. Herman was still looking for finbacks quite a ways offshore. But, he was optimistic that he would bring a huge bounty home as well. When the *Dennis Gayle* arrived in Fields Landing, a crowd had formed to see the first whale of the season hauled in. As the boat eased up toward the 25-foot-wide whale slip, a rowboat carried the end of a steel cable out from the station. Once the cable was made fast to the whale's tail, or fluke, the boat crew released its lines from the whale and the animal was pulled to the end of the slip by a big steam winch.

The whale was then pulled up the slip to the main deck for flensing and processing while hundreds of tourists watched. It only took about two hours for the whale to be completely flensed. Then, the land crew would clean and rinse off the decks and make ready for another catch. For the first venture of the inexperienced whaling captain, he was very pleased with his adventure.

Skipper Gib with son Dennis and daughter Gayle at Fields Landing Whaling Station

Chapter Nine

GOOD AND BAD FORTUNES • 1948-1951

In June of 1948, Gib and Maxine were blessed with the birth of their third child, a boy, who would later become an integral part of his fishing business. They named him Bill, after Gib's brother Bill and his father Will.

The year was also blessed for America's only operating whaling industry. Regulations allowed the harvesting of 100 whales a year by each boat and both Gib and Herman were doing very well. But then, fate dealt them a bad hand!

The *Dennis Gayle* was several miles out to sea headed in because of bad weather when Gib looked toward land and spotted smoke. A lot of smoke coming from the direction of Fields Landing, the site of the whaling station. Fire had encompassed the small station.

The *Lynnann* was just docking. Herman called on the radio: "*Dennis Gayle*! *Dennis Gayle*! You'd better get in here. The station's on fire. Over." Gib came back "Herman. We'll be there in a couple of hours! *Dennis Gayle* clear."

When Gib landed, he met Herman at the floating dock, just north of the whale slip. Nothing of the whaling station existed. Nothing! It was like it never existed except for the smoldering embers and whale's teeth scattered all over the barren landscape.

Fortunately, even though several workers were there at the time, no one was injured. But, both boats and their crews remained docked while repairs began. The townsfolk of Fields Landing rejoiced because they no longer had to put up with the smells. American whaling was finished!

In the meantime, a fish processing plant just north lie dormant. Gib and Herman figured they needed to keep their boats busy while the whaling station remained closed. So, they equipped their vessels for bottom fishing and formed a fish processing plant. Gib's dad, Will, moved down from Waldport to help run the plant. Gib's brothers, Bill and Glenn, brought their fishing boat, the *Pearl Harbor* down from Newport to deliver fish to this new plant. Over the next few years, the processing plant began to grow and prosper as more fishing boats delivered their catches.

By 1951, repairs had been made to the whaling station and the two skippers had to make a decision. Do they continue fishing? Or, do they go whaling? They decided to do both. The whaling season was usually from May through October. Their two boats would go whaling during that six months, then fish the remainder of the year. They hired a general manager to run the fish processing plant while they divided their time between fishing and whaling.

Gib's love was the sea. And, whaling was his calling. Once again, the whaling season began along the Northern California coast, Gib and Herman would start friendly competition to see who could bring in the most whales during the season. One day, Gib set a course to Punta Gorda, just south of Cape Mendocino, where he heard from fishermen that a pod of humpbacks were blowing. On the slow trek south, they searched for hours. Then, they noticed that dolphins came to jump right in front of the boat and a flock of seagulls flew by. Gib said "signs are good that feed is nearby and whales might be near. Maybe those dolphins are leading us right to 'em. I'm going down to the wheelhouse to call Herman and see how he's doin'." He called the *Lynnann* and Herman answered "I'm off the Cape, Gib. Nothin' here at all! *Lynnann* out!" Gib thought that was a pretty abrupt radio call. He then smiled, as he realized that Herman must be in the middle of a bunch of whales. Up on the bridge, he said "Herman's chasing some. Bastard isn't helping at all. We're on our own. I think we'll find some……." Just then, a crewman cried out from the crow's nest, "Gib! I see a bunch of blows off to the southwest! Looks like our pod of humpbacks!" At the end of the day, the *Dennis Gayle* was towing two big humpbacks back north to Fields Landing. And, Herman? He, too, had a couple of finbacks and his *Lynnann* was heading in as well. The small whaling station was about to get very busy.

All seemed to be going well at sea, but on land, it was a different story. The townspeople started to get bitter over the overwhelming stench from the whaling station. One day, when Gib unloaded a huge 63-foot finback whale, a mob appeared and was waiting for him. He called to his crew "Hey, guys! Be careful. I'm not sure what this bunch wants, but I'll deal with 'em." But Gib's crew members would not let their skipper go this alone. So, as a group, they marched to face the unruly mob.

"Stop this stench or we'll do somethin' about it" one angry member of the mob shouted! Two police cars showed up and prevented anything physical from happening. Skipper Gib offered to meet with the crowd to try to come up with a solution. They reluctantly agreed. That same night at the meeting, following much give and take, Gib came up with a solution. He called his whaling station crew together and said "I want daily scrubbing of these decks and whaling slips to make sure we cut down on the stench. We'll have daily inspections to make sure we take care of these people here in town."

So, for a while, the townsfolk of Fields Landing seemed pleased with the results. But then, on a foggy summer day in 1951, another fire erupted and gutted the entire whaling station and several nearby houses. This fire was much more destructive than the first one, with toxic black smoke spewing into the air. It took the fire crews hours to finally put the fire out. Gib and Herman always wondered if this fire was intentionally set, but they could never prove it. So, they made a decision: the whaling station in Fields Landing was finished and so was whaling off the Northern California coast.

Chapter Ten

PRESENT DAY

The old skipper never saw rogue waves like this in all his seventy-five years at sea! He knew his small boat couldn't take much more of this merciless beating! He still couldn't get any response from his crew down below and it was all he could do to keep the *Winga* from capsizing! He tried hailing the Coast Guard one more time, but his reliable radio shorted out. Now he had no way of contacting anyone.

He was alone with only his savvy and a couple of flares to bail him out! All of a sudden, another huge swell attacked from straight off the bow. Gib grabbed the wheel and held it with all his might. He increased speed and the boat shot straight up in the air. As it hit the top of the wave, he let off on the power and the *Winga* began its long 30-foot free fall smashing into the next wave.

Suddenly, Gib thought he saw a flashing light dead ahead off in the distance. His instincts told him to reverse course and head toward land. But the light seemed to beckon to him so Gib headed his boat straight for it.

Chapter Eleven

HEADED FOR
SOUTH AMERICAN WATERS • 1952-1954

For a time, skippers Gib Hunter and Herman Foland converted their whalers to fishing boats, fishing for sole and rock cod off the Humboldt coast. The fish plant was expanding and Gib now had purchased several used fishing boats to add to an ever growing fleet. But, the call to go whaling was very strong, particularly for Herman. One morning at the breakfast table, while sharing precious time with his wife Helen and his daughters Carol and Beverly, Herman's mind was elsewhere. Helen was used to it. She knew Herman was longing for the pitching deck of the *Lynnann* out at sea. Just then, the phone rang. Herman answered on the first ring. "Hello. Foland speaking," he said. The tinny voice at the other end said "This Herman Foland, the whaling captain?" "Yes, it is. Who's this?" "My name's Raul Ruelas." I understand you have a whaling boat." "Yes, I do", Herman replied. "I'd like to set up a meeting. I can be in Eureka in two days", the caller said. "We need a whaler in South America and you could be the man we're looking for." After a quick glance at Helen, Herman began to smile. He said into the phone "you name the time and place. I'll be there."

Two days later, the meeting took place in in the same lounge that Gib and Herman made their first decision to go whaling. Herman arrived a few minutes early. He chose a table in the back of the room for privacy, ordered a cup of black coffee and waited. He was about to give up when two men with tan suits and top hats entered the lounge. They glanced in Herman's direction, made eye contact and headed to the back table.

As they approached, the small but stocky man introduced himself as Raul Ruelas and his partner as Angel Fargas. They sat down across from Herman and explained that they owned a small whaling station in Lima, Peru and were in search of whaling boats and crews.

Herman considered this and did the math. Lima, Peru was over four thousand nautical miles southeast of Humboldt Bay. His boat, the *Lynnann* could travel at eight to ten knots an hour. It would take three solid weeks just to get to Peru, not counting several fueling stops. But the future adventure started to stir in his mind. He decided to consult with his partner, Gib, who just came in from a successful fishing trip. They met at the dock and decided to go inside for a cup of coffee.

Herman told Gib of the offer made by Ruelas and Fargas. Herman was itching to head south, but Gib was more hesitant. Finally, the decision was made. Herman would take his boat to Peru while Gib would stay behind and manage their fish business.

So, the preparations began. But, Herman figured "why prepare one boat when you can prepare two!" Gib decided to send his boat and crew, the *Allen Cody* to Peru with Herman. To make that long journey, the *Lynnann* and *Allen Cody* needed some extra fuel tanks. Finally, after two months of getting the boats ready for sea, the *Lynnann* and *Allen Cody* crossed the Humboldt Bay Bar side by side, turned south and started the long 21 day journey to Peru while Gib stayed behind to fish his boat, the *Dennis Gayle* and manage his seven-boat fishing fleet.

The Pacific Ocean showed many personalities on the journey to Peruvian waters. North of the Equator, it was the beginning of summer which presented many foggy days and smooth waters. But, in crossing the

Equator, winter began and Mother Nature started to show her fury. The *Allen Cody*, skippered by a crusty veteran named Bud Newton, trailed the *Lynnann* by ten miles and the boat crews couldn't see each other over the horizon. They were entering a hurricane zone and had to slow their speed to four knots. Powerful swells pounded the boats and all the crews could do was hang on for dear lives! They were so far offshore that changing course to find shelter was out of the question.

Lady luck seemed to be riding with them. After four days of taking all the Pacific could deliver, they journeyed into calmer waters off the coast of Ecuador. Herman told both boat crews to "crank it up to eight knots"!

After twenty-two days, the *Lynnann* finally turned inland toward Callaro, Peru, a seaport adjacent to the capital city of Lima. He called Captain Newton on the *Allen Cody*, and told him to pick up speed and join him for the short jaunt to Callaro. As they were headed in, they could see whales blowing all around them. Herman said to one of the crewmen "this will be a great trip! They're plenty of whales here!"

After contacting Peruvian customs officers, they were given permission to dock. A small tug pulled alongside and guided the *Lynnann* and *Allen Cody* to the dock. Raul Ruelas and Angel Fargas were waiting for them. Skipper Herman had a surprise waiting for him as well. His wife Helen and daughters, Carol and Beverly, were on the dock with big welcome signs that made both crews excited. They had flown down from Northern California and had rented a house in Lima.

Herman, his family and crews, settled into the winter months of Peru. His boats were regularly bringing in whales. He'd often come home and tell Helen and his daughters about whaling adventures at sea. In the meantime, Gib kept his boats busy fishing and he was expanding their seafood processing plants along the California coast.

Chapter Twelve

WHALING BEGINS AGAIN IN NORTHERN CALIFORNIA • 1958-1959

One summer day when Skipper Gib was home playing catch with his two sons, he didn't realize his life was about to take another dramatic turn. Just as one of the boys was about to throw him the perfect pitch, the phone rang and Gib started to run into the house to answer. The boy, not knowing that Gib was headed for the house, threw a perfect strike to where Gib was…and the ball headed straight for and through the living room window. Gib ignored the crash when he got the message on the phone. "Gib", the caller said "My name is John Caito. I'm calling from Richmond, California in the San Francisco Bay area. We're setting up a whaling operation. I understand you have a couple of whalers." Gib responded "I've had enough of running a whaling station. Didn't turn out too well up here in Fields Landing as you know. Damn place burnt down twice!" The caller said "Gib. Wait a minute! I'll run the station. I just need you and your boats to go out and catch us a few. The season will run from May to the end of October each year. You can go back to fishing for the remaining six months. How soon can you get down here?" Gib knew a good opportunity when he heard one. He knew Caito had a good reputation in the fish business. He told Caito he'd call him back the next day with a decision.

That night, at the dinner table, Gib and Maxine, with their three children eavesdropping, talked about their future. Maxine pleaded with Gib to stay the course and run the successful fish processing plant he had built in Eureka. In fact, his plant has receiving stations up and down the Oregon and California coastline. He had plenty to keep him busy with his plant and fishing boat fleet. By staying in Eureka, he could also spend more time with his family. His three children were growing up. His oldest son was a sophomore in high school; his daughter was in junior high and his youngest son was only nine years old.

But, whaling was in Gib's blood, so the decision was made. His General Manager would run the fish processing plants while Gib would spend six months out of the year hunting whales off the San Francisco coast. He called Caito the next day and told them his decision: he and his boats would be ready by the start of the next whaling season.

By the time Gib was ready to head down to Richmond, partner Herman Foland had had enough of whaling in Peru. He was ready to head home. He, too, would join the whaling fleet off the California coast. In May of 1958, the newest and last era of American whaling began with five boats and 25 crew members.

Strands of fog wisped through the straits of San Francisco Bay as the five whale boats headed for sea. Rounding the corner at Pt. Richmond past the Brothers lighthouse, their course took them close to Angel Island and under the Golden Gate Bridge outbound for the open Pacific waters. On the *Dennis Gayle*, Gib Hunter set his path directly toward the Farallon Islands just 30 miles to the west of the San Francisco coast. By studying various nautical charts, Gib discovered possible feeding grounds for the Humpback whales around the islands in what is called the Gulf of the Farallons.

Skipper Gib managed to recruit a very experienced harpooner named Eric Neilsen from Norway. Neilsen was a short, gray-haired gunner with a weathered face that always seemed to wear a constant smile.

Once they sailed under the Golden Gate and watched Point Bonita pass on their starboard side, the *Dennis Gayle* crew anticipated a great adventure. They looked up and saw an albatross fly overhead and hoped it didn't portend a bad omen. But, it did.

A good sized humpback was spotted about fifteen miles offshore. Gib steered the boat toward it and the excitement began. Gunner Eric loaded the gunpowder into the harpoon tip. He shouldn't have been smoking a cigarette then. It accidentally sparked the gunpowder and the harpoon tip exploded. When the smoke cleared, Eric yelled for help. The crew rushed to his aid and led him inside. His right hand was pouring blood. Gib grabbed a towel and wrapped his hand. He told Angelo Torino, one of his crew members, to apply pressure.

Gib ran upstairs to the wheelhouse where Ernie Jones, another new crew member, was standing watch. Gib grabbed the radio and called into the whaling station to tell them the boat was heading in with an injured crewman. It took a long 90 minutes for the *Dennis Gayle* to backtrack toward Pt. Richmond and tie up at the dock. An ambulance was waiting. The EMT complimented the crew for saving Eric's life by keeping pressure on the wound, yet they were not able to save two fingers of Eric's right hand.

From his hospital bed, Eric Neilsen told the doctor to "patch me up. I need to get ready to go to sea. We have whales to catch!" Skipper Gib and several of the crew spent hours visiting their wounded gunner. Gib told Eric to "take your time. We're not going anywhere. Weather's too bad. No one's going out there."

Finally, two weeks later, the boat and its crew headed out for their second maiden voyage. This time, it paid off. A huge pod of humpbacks was spotted close to the big South Farallon Island. It seemed the whales were playing around with the boat and kept outsmarting its skipper. Every time Captain Gib maneuvered the boat to where he thought the whales would surface, they would appear several yards behind their stern. So the chase was on. After several hours with the sun starting to set, the boat slowed to a stop. Everything fell silent and several sets of eyes started to scan the horizons. Eric yelled "here they are, right in front of us!" Gib pulled the throttle to full stop and the wait began. The whales seemed to rest right below the surface, but Eric could see them. All of a sudden, a big beast gradually came to the surface and blew its steamy breath into the sky. Eric didn't hesitate. He slammed his bandaged hand against the cannon's trigger and…BOOM! A great shot! The harpoon struck the whale right below its dorsal fin. But, the harpoon grenade failed to explode. Without the explosion, the harpoon was like a fishing line holding onto a fighting trout. The whale weighed almost forty tons and was not happy! Gib advanced the throttle to keep up with the whale as Eric and members of the crew prepared the gun for another shot. Fathoms of cable began to spin out of the bow of the boat as the humpback headed for the deep. Ernie Jones, the new member of the crew, kept water on the cable winch so it wouldn't heat up and snap the cable while it continued to unspool. Eventually, the whale began to tire and head back to the surface. The plan was to get the whale to surface right in front of the boat which would allow another shot. As the whale surfaced, Eric again pushed the trigger and this shot was successful. The *Dennis Gayle* had its first forty-foot humpback of the 1958 inaugural season.

 With the whale successfully tied alongside, they headed for Pt. Richmond to unload their big catch. Under the Golden Gate Bridge, they discovered the currents caused by the outgoing tide were almost too much for the boat and its heavy cargo to bear. Their usual 8-knot speed was reduced to around 4 knots and it took forever to navigate through San Francisco Bay passing Angel Island, under the Richmond-San Rafael Bridge and around the corner to Pt. Richmond. But, they made it. The whaling station owner, John Caito, was standing on the dock with a cigar in his mouth. "Way to go, Gib!" he yelled. "That's a beauty! Also, I just heard from Herman. He's on his way in with a big finback so we're off and running!" The first whaling season was a great success as each boat reached its quota and no further injuries occurred.

Gib Hunter and his crew in the Pt. Richmond Whaling Station

News of the only American whaling operation reached the East Coast and reached a television game show producer who had a great idea. "What if I could get an American whaling Captain to appear on our game show *To Tell the Truth*? No one could guess who this guy is." So the connection was made. Captain Gib Hunter and his wife Maxine flew to New York and Gib actually appeared on *To Tell the Truth*. Entering the CBS Studios and seeing the bright lights of the set, Captain Gib was somewhat intimidated. He looked over and recognized the host Bill Collyer and panelists Kitty Carlysle, Tom Postem, Gig Young and Polly Bergan. "What am I doing here?" he thought as he sat all dressed in a coat and tie amid the hot lights. "I belong at sea, not in front of these TV cameras." But once the show started, Gib settled in and was a great contestant. The other three contestants were very convincing and actually looked like whaling Captains while Gib looked like a businessman. He seemed to have all the panelists on the show fooled. And, to tell the truth, no one, except Gig Young, made the correct guess of who the whaling captain was. So, one of the last American whaling captains became nationally famous!

Chapter Thirteen

YOUNG SONS COME ABOARD

During the next season, Captain Gib Hunter decided to bring his young sons on the boat during the summer to learn the trade. Sixteen-year-old Dennis and ten-year-old Bill were excited to be part of the crew. The boys were given their own bunk room right across the engine room and were told not to run around and get in the way. It didn't take them long to get acclimated. They spent their first night on the boat at the dock as Captain Gib planned on getting underway just before dawn. That would put them at sea by daylight. It was hard for the boys to get to sleep as they were so excited about their first whaling voyage.

All of a sudden, Dennis and Bill woke to the loud sounds of the huge diesel engine starting and the sudden vibrations of the *Dennis Gayle* getting underway. They quickly dressed and headed for the bow. The view of the big Richmond-San Rafael Bridge was amazing! Bill said "Gee! Hope Dad doesn't hit the bridge!" But, as the whaling boat sailed under the bridge unscathed, Bill felt a little better.

Captain Gib again headed for the Farallons. After listening to fishing boats talking on the radio, he knew a large pod of finbacks were seen about 20 miles west of the Farallons so he set a course due West! There was quite a swell building, caused by a Pacific storm much further out to sea. The boys were instructed to hang on and be careful as they headed out on the eight-hour journey to the whaling grounds. Once they were underway, they noticed the smell of breakfast being prepared in the galley by the boat's cook, Rudy Vaverca. Rudy yelled "C'mon boys. Breakfast is ready." What a breakfast! Pancakes, eggs, bacon! Wow! This was going to be a great adventure! They had just finished breakfast and started to leave the galley when Rudy said "Hey boys! Where ya' going? You got to work on this boat. These dishes need to be washed and dried. Here ya' go!"

Young Bill Hunter and his dad Captain Gib on the bridge

After the boys finished their first task of washing and drying the dishes, the crew settled into card games in the galley while Gib ate breakfast in the wheel house.

After an hour or so, Dennis started to feel a little queasy. "Shouldn't have had all those pancakes", he was thinking. He made his way to the back deck and started to lean over the stern to deposit his breakfast into the sea. Quietly, younger brother Bill made his way onto the back deck and stood right behind him. "Whatcha' doin', bro?" Bill said. "Feeding the fish?" Crew members started to assemble on the deck, heard the conversation and started to laugh. Finally, Angelo Torino took Dennis by the arm and led him to the bunk room. He confided "that most crew members get seasick on their first voyage, so don't worry. Just lay here for a bit. If we see any whales, I'll come and get you!"

After a few hours, Dennis started to feel better and began to get used to the boat "rockin' and rollin'". Captain Gib called to him and told him to get his brother and come up to the wheel house. Dennis and Bill climbed the stairs, holding on while the boat pitched up and down. Entering the wheel house, Gib told them "it's time you learn how to take a wheel watch!" He showed them the compass which was set at 270 degrees. "Make sure we keep on that course", Gib said. Then, Dennis noticed the big steering wheel turning from side to side on its own. It seemed to make groaning noises as it turned. "What's that?" Dennis asked. "Shouldn't we grab the wheel? What's that groaning noise?" Gib said "that groaning sound is from Iron Mike here. It keeps the boat headed straight on course. If we need to change course, we reach for this switch". Gib pointed to the on-and- off switch for the automatic pilot named Iron Mike. Gib continued "all you need to do is to sit here, make sure we don't hit anything and check that compass. I want it on 270 degrees. Dennis. You sit in this chair and Bill, you stand over here". Dennis sat in the seat while their dad put a small Captain's hat on both the boys to celebrate their first wheel watch. "I'll be back in a few minutes", said Gib. He then left the wheel house and the boys glanced at the open ocean ahead. Bill gasped and said "Hey bro! Hope we don't see any of those Russian subs come up in front of us!" No subs surfaced, but white caps started to form as the Pacific storm began to move in. The boat started to pitch more and more. But the Iron Mike kept the compass right on 270 degrees.

Chapter Fourteen

PRESENT DAY

The old sea Captain kept his course headed toward that flashing light. If his compass stayed on 270 degrees, he could find out what that light was. Maybe it was a freighter or tanker. If he could get their attention with one of his remaining flares, perhaps they could rescue him and his crew, wherever they were. It was strange the crew didn't answer. He was too busy to worry about the crew because it was all the old man could do keeping on course and maneuvering through rough ocean waters to keep the small boat from capsizing. Then, with a clap of thunder and bright lightning, Mother Nature wreaked her fury! Torrential rains came and the old skipper could hardly see ahead of him. His only hope to save his boat and his crew was to catch the attention of the passing ship. The flashing lights were still in front of him, but they were getting harder to see. He opened the door and headed to the port side. The rain and wind blasted his face, making it harder to see. He couldn't take the chance on using his remaining two flares. No one would see them with all the lightening around. Then, without warning, another rogue wave came.

Chapter Fifteen

YOUNGSTERS' WHALING ADVENTURES CONTINUE

While Dennis and Bill were up in the wheel house, the skipper called for a crew meeting in the galley. Gib said "I don't like the way this storm is headed. I think we should head for the South Farallon Island and anchor up." The Farallon Island chain lies 30 miles outside the Golden Gate and 20 miles south of Point Reyes. The South Island is the largest in the chain, pyramidal in shape. Its peak, called "Tower Hill", is 357 feet high. This would provide the shelter this 100-foot whale boat would need.

Gib headed for the wheel house and congratulated the boys for a great wheel watch. "I'll take it from here. Why don't you go down to the galley? I think Rudy has some hot cocoa ready for you." The boys made their way downstairs while Gib reversed course and headed for the Farallons.

As the boat slowed and rounded the corner behind the shadows of this 95-acre island, the violent pitching stopped and it seemed like the boat was floating on a calm lake. The crew was on the bow ready to drop anchor with Gib on the bridge. As the boat drifted to a stop, Gib yelled "drop the anchor!" The sounds of the heavy chain filled the air as the anchor headed for the bottom of the sea. As the anchor caught on the rocks below, the *Dennis Gayle* was anchored and protected from the storm. They finished anchoring just as the sun set behind the majestic Tower Hill peak and it seemed like the temperature dropped twenty degrees. The crew bundled up and prepared the boat for a cold night.

As one day stretched into two, the storm was relentless around them and the crew spent their time maintaining equipment and playing cribbage in the galley. The boys, on the other hand, were getting bored. Angelo had an idea. It's time to play a practical joke on the boys.

Late one afternoon, he went to the boys and told them to get dressed up. He would take them to a drive-in movie on the island. A top Western was playing! Dennis and Bill got excited. "Really! How do we get there?" Angelo said, "We'll take our small skiff and row there. Hurry now and get dressed. I want to get there before dark." The boys hurried to their bunk room and got dressed for the big movie. "Gee, bro", said Bill. "I'll be so glad to get off the boat for a while. This'll be fun!" The crew was waiting in the galley for the boys to show. They weren't disappointed. The bunk room door opened and out came the rookie crew members all spruced up and dressed in their Sunday best! As the boys walked down the hall, Captain Gib came down from the wheel house and saw them. "What the hell are you doing dressed up like that?" Dennis said "Gee, dad. Angelo is taking us to the drive-in movie on the island!" Gib was taken aback! All of a sudden, the crew members came around the corner and began to laugh.

Gib told the boys that there was nothing on the island but a bunch of birds. "And, we're not disturbing them. Go in and take those damn' clothes off and get back into your coveralls." As the disappointed boys slowly made their way back into the bunk room, Gib looked at his crew, then paused, then flashed a grin. "At least, it made the boys clean up. They were starting to smell" he said with a laugh.

The next day didn't show much promise either, so the *Dennis Gayle* sat still on the lee side of the islands. The boys were really looking for something to do. "I thought whaling would be more exciting", Bill said. In the early afternoon, the boys noticed a flock of seagulls flying low looking for food. Bill and Dennis went into the galley and got some pieces of bacon. They tied these pieces to a rope. Then, somehow they got an idea! "Let's put some tabasco sauce on it", said Dennis. So they did. They spent some time watching the gulls take the bacon and then head toward the water to wash off the tabasco. After hearing all the laughter coming from the stern, the skipper thought he'd better check on the boys. "Oh my God!" Gib exclaimed. "Would you like someone doing that to you? Knock it off!" After a slight pause, the skipper said "If you're looking for something to do, go down into the engine room and polish all that brass." Obediently, the boys spent several hours deep in the engine room. When they completed their job, they headed for their bunks and fell fast asleep.

After another day of tedium and boredom, the storm wasn't moving. It was apparent they were going to have to head in to Richmond without a whale alongside.

The boys were excited they were going to finally get "off this bucket". The crew pulled anchor and the *Dennis Gayle* headed in to San Francisco Bay. The sea was swirling but the swells were actually pushing them toward port so the thirty-mile journey wasn't too bad. Four hours later, after sailing under the Golden Gate, Gib turned the boat a few degrees to port, headed alongside Angel Island, then under the Richmond-San Rafael Bridge and the safety of San Pablo Bay.

So ended the first whaling adventure for Dennis and Bill.

As the years passed, the whaling seasons were very successful. All the boats were meeting their quotas during the six month seasons and during the winters the boats would be equipped with drag fishing gear and would be off the Humboldt Bay coast catching sole and cod to deliver to their Fields Landing fish processing plant. On the whaling front, Eric the gunner finally retired and Gib recruited an experienced gunner from Japan named Ken Hamai.

Dennis attended high school and would work on the whaling boat during the summer. One whaling trip he would remember for the rest of his life. The trip started in the middle of summer with the *Dennis Gayle* heading for a pod of finbacks. Captain Gib told Dennis to climb up into the crow's nest and start watching. "If you see a blow, yell down and tell us!"

Dennis started his climb up the rope ladder into the crow's nest. It was scary trying to hold onto the ladder while the boat was rolling back and forth. "God! Don't look down", he thought. Dennis could visualize a 100-foot fall into the churning Pacific Ocean and knew that would be fatal. So, he hung on for dear life!

Finally, he made his way into the crow's nest which was more comfortable than he anticipated. *What a view!* he thought as he could scan 360 degrees to a clear horizon. As the hours passed, boredom set in. *Good thing I brought my transistor radio* he said to himself. As the music of Elvis Presley's *Don't Be Cruel* played in his ears, he started to get sleepy. The warm air and the gentle rocking started to get the best of him and he slumped over in peaceful sleep. Little did he know that the Captain was making his way up to the crow's nest! Gently touching the boy's shoulders, Gib yelled "what the hell you doing? You can't spot a whale with your eyes closed! And, turn that damn Elvis music off!" Startled, Dennis woke up and resumed his scan.

Gib made his way back down to the bridge, smiling at the crew members. "That'll teach the kid a lesson", he said. All of a sudden, a cry from the crow's nest could be heard. "I see one! I see one!" Dennis yelled. "Right over there" he said as he pointed to the port side. "By God, the kid spotted one!" Gib said proudly as he turned the wheel toward the port side. As the boat turned, Dennis started to climb down from the crow's nest. The boat rolled dramatically and it was all the kid could do is hang on! But, he made it to the bridge.

Gib and his crew on the bridge of the **Dennis Gayle**

Now, the chase was on. The boat slowly started to close in on the big finback. What a beautiful beast! Looked like it would measure over 60 feet. Just as the boat drifted near, the big whale showed its flukes and headed for the deep. The stop watch started and the guessing game began. Gib estimated that in 15 minutes the whale would head for the surface. He told the crew members to be ready. Ken Hamai, on the bow, got into position and was ready to shoot. Just about right on time, he could see the whale coming to the surface. "Gib. Stop the boat! He's coming up right in front of us!" The Captain pulled the throttle to full stop and the boat began to drift. Then, one hundred feet right in front of them, the whale lazily made its way to the surface and began to blow. Ken didn't waste an opportunity.

He slammed his hand on the trigger and the harpoon began to fly toward the big finback. Bam! A direct bullseye! As the harpoon entered the whale, the grenade on the tip failed to explode. Like a horse grabbing the bit in his teeth, the whale began to take off due west! A second shot would have to be made to humanely subdue and kill the whale. Gib had an idea!

He looked at Dennis and asked "Would you like to take the second shot?" "Me?" Dennis replied. "Yes you!" Gib said. "Get your butt down on the bow. Hey Ken! Dennis is coming down to take the second shot!"

Dennis quickly made his way down the stairs and across the catwalk to the bow. Ken greeted him with a big smile and gave his instructions: "Okay. Here's what you do. You watch the whale come to the surface. When he noses up and starts to breach, you watch for his dorsal fin. That's your target. Don't shoot too early. Just watch for the fin. When you see it, slam on the trigger. That's all there is to it. I'll stand right here beside you. Now, get ready!" Dennis took his position behind the big cannon, his hand on the trigger, ready for action. All of a sudden, he could see the big finback right below the surface. He was coming up. "Oh, God! Please don't let me miss", said Dennis. "Here he comes" yelled Ken.

As the whale came to the surface, his nose broke through and Dennis fired! He shot too early and the harpoon bounced off the whale's nose and vaulted into the air. "Damn!" yelled Captain Gib from the bridge. "Dennis. Get your ass off the bow and get back up here now!" So, a disappointed young crew member slinked off the bow and back up on the bridge with the memory of missing his first and only shot at a whale etched forever in his mind. Oh, yes. They finally caught that big finback. It was sixty-five feet in length and weighed over sixty tons. In spite of the miss, the trip turned out to be a great success. On the way in, Ken consoled the young crew-member by telling him of his first whaling experience in Japan. "We were lined up on a huge finback, I took the shot and missed. The damn whale dove and we never saw it again."

Later that summer, word spread that a large pod of finbacks were feeding about twelve miles west of the Farallons. The boats left the dock just after midnight and steamed toward their prey. As the sun was rising in the east, the crew began its search. On the way out to sea, gunner Ken gave a quick lesson on how to spot different types of whales. He told Dennis "when you see a big high thin blow straight up in the air that could be a big blue whale or a finback. They can blow at least fifteen feet in the air. A humpback also blows straight up in the air, but its blow is much thicker. A sperm blows at a 45 degree angle." Skipper Gib, overhearing this instruction, told his young son to get up in the crow's nest and use his new knowledge. As the boat continued west, the water turned a deep blue. It should be easy to spot a whale. Hours later, as Dennis was scanning the horizon, he saw this tall column of steam. He wanted to be sure, so he waited. There it was again! He yelled "I got one! Straight ahead on the horizon!" By that time, the skipper and the rest of the crew spotted it too. Gib cranked the throttle to full and the race was on.

As they got closer to this slow moving whale, Ken yelled "I think that could be a big blue whale!" The blue whale is the largest living mammal on the face of the earth. This particular whale seemed like it was just lying there sunning itself. The boat drew near and Ken was ready on the bow. The whale was skimming along near the surface and took another blow. Then, it readied to dive. As it arched its huge back, Ken was ready. He fired! As the smoke cleared, everyone could see he had a clean shot. But, the grenade failed to explode. Now, they had a mad blue whale charging out to sea. It was all Gib could do to keep the beast from snapping the cable. All of a sudden, the whale turned and headed for the 100-foot boat. It kept coming and coming! It smashed into the port bow driving the harpoon out of its body. The crew never saw the whale again. After retrieving the harpoon, the skipper headed to the bow to check for damage. The entire port section of the bow was smashed in, but there was no danger of sinking. It would take the *Dennis Gayle* out of action though. Gib headed back toward the bridge and turned the boat for home. It would spend the next few weeks on dry dock getting the damage caused by the big blue whale repaired.

Chapter Sixteen

HISTORIC FLOOD AND WHALING TAKES A SEVERE BLOW • 1964-1971

As the whaling season ended, Gib brought his boat north to fish along the Humboldt coast during the winter months. Little did he know what this particular winter had in store!

Starting in late December of 1964 through the early part of January 1965, intense rainfall hit the West Coast hard. Over 22 inches fell on Northern California's Eel River basin in a span of only two days. The system was referred to as the "Pineapple Express" funneling water and warm air which melted record snow packs. Water rushed down the Eel and neighboring Van Duzen Rivers. To the north, waters from the Mad and the Smith Rivers also were rushing into the sea. The mammoth flood killed nineteen people and heavily damaged or completely devastated almost a dozen towns. The floodwaters destroyed twenty major highway and county bridges and carried away millions of board feet of lumber and logs from mills sites. Many communities suffered massive power outages. In what was termed as a "one-hundred-year" flood, the Humboldt Bay area was literally cut off from the rest of the world.

After the rains subsided late in January, the communities started to rally, but the big crisis they faced was getting food and supplies into the area. The only transportation lanes were by sea and air as land transportation north, south and east was non-existent.

Among the volunteers heeding the call for help was a whaling Captain and his two crews. Gib volunteered his whale boats, the *Dennis Gayle* and *Allen Cody* to transport food and supplies from the Bay Area to Humboldt Bay. Round trips would take at least two days depending on the weather. The 31-hour trips one way to the Bay Area were treacherous as Mother Nature continued to pound the west coast of California. Fortunately, the two converted Navy Inter Island cargo boats had plenty of storage space so they could bring plenty of supplies back to Humboldt Bay. On return trips north, the two boats had to battle heavy seas and strong head winds. The wooden hulled boats were sturdily built and could withstand the pounding the Pacific would dish out. Several months later, thanks to hundreds of volunteers and Mother Nature calming down, the north coast of California began to dry out and return to its new normal.

As the months and years rolled on, the six-month fishing seasons and six month whaling seasons were very fruitful as quotas were reached and the crews were happy. Up in Eureka, Gib's youngest son, Bill, began to take an active role in the fishing industry and oldest son, Dennis, went off in a different direction. In 1969, Captain Gib's reputation as an innovative whaling and fishing captain began to spread. He was asked to do some research on deep and mid-water fisheries. At the same time, Gib heard of a former Coast Guard cutter that was just taken out of commission in Southern California. Ironically, this cutter named the "Ewing" spent some time patrolling the waters off the Humboldt coast. Gib figured this boat would fit into his plans nicely: it could hire out for research for six months out of the year from November through April. The May to October period could also be put to good use. The 135-foot vessel would haul whales for the *Dennis Gayle* and *Allen Cody* so the hunter boats wouldn't have to go back and forth from sea to Pt. Richmond. The boat was named the *Pacific Raider* since Gib was a passionate fan of the Oakland Raiders.

*The **Pacific Raider** at the Eureka Fisheries dock in Fields Landing*

All was going well until the United States government decided to put three species of whales on the endangered list – the humpback, blue and gray. At first, domestic hunting of these whales was allowed, but pressure began to mount. In 1971, the government prohibited domestic hunting of all whales, however, the importing of whale products would still be allowed.

Gib and his crew were flabbergasted! They had been a part of the whaling industry since the late 1940's and now they were shut down. Dennis and Bill made their way to Richmond to experience the last whaling adventure of their lives. The "Dennis Gayle" captain and crew planned on heading out to sea at dawn. The night before, Captain Gib listened to the weather report calling for heavy seas and high winds. There was no way the boats could function in 30-foot seas. So, the colorful, rich history of American whaling ended in a whimper. John Caito, the owner of the whaling station said in a March 2, 1971 article in the *Richmond Independent*: "This lays to rest three hundred years of tradition. There will be no more Ishmaels, no more Ahabs, no men to carry on the traditions of New Bedford or Nantucket." He continued "This will have no effect on the whale population. The largest whalers in the world are the Russians and the Japanese, and this won't affect them at all."

It turned out he was right. The five U.S. whalers were prevented from taking their quota of 42 finbacks, 75 sperm whales and 52 sei whales. The foreign whaling factory ships from Norway and Japan moved into the fertile whaling grounds off the California coast and literally wiped out the Pacific whaling population. The five whalers headed north to Humboldt Bay and converted to enter the commercial drag fishing business. All the harpoons and harpoon guns were disassembled, taken off the boats and put into storage. The era of whaling off the California coast was finished and would never return!

Chapter Seventeen

PRESENT DAY

The rogue waves came in bunches now. It was only a matter of moments before the small fishing boat would be smashed to bits in the horrendous storm. Gib could see nothing but water all around him. In the 75 years he'd been at sea, he'd never faced a crisis like this one. But, where is the crew? "I've got to get them ready to abandon ship", Gib thought. "And, where's that damn freighter I saw right in front of me." Then, *wait a minute*, he thought. *There's that damn flashing light dead ahead. Kinda reminds me of the buoys right off Newport.*

He flashed back to the days when whaling was an industry of the past - to the days he pioneered new fishing methods. He was proud of the crews and the boats that had developed quite a fishing industry along the West Coast. He knew that his younger son Bill and grandson Travis would follow in his footsteps in the commercial fishing industry.

If I don't make it through this, I've lived quite a life. He thought back to his 80th birthday when his whole family surprised him with a huge party and his son sang the song *My Way* which reflected the way Gib lived his life. *You know*, he said to the lonely wheelhouse, *I did do it my way.*

The light got closer now and he somehow knew that the heavens were reaching out. As the small boat *Winga* and its Captain finally lost the battle and succumbed to the sea, Gib reflected on the poem by Alfred Lord Tennyson:

CROSSING THE BAR

Sunset and evening star,
    And one clear call for me!
And may there be no moaning of the bar,
    When I put out to sea,

But such a tide as moving seems asleep,
    Too full for sound and foam,
When that which drew from out the boundless deep
    Turns again home.

Twilight and evening bell,
And after that the dark!
And may there be no sadness of farewell,
    When I embark;

For tho' from out our bourne of Time and Place
    The flood may bear me far,
I hope to see my Pilot face to face
    When I have crossed the bar.

THE END

AFTERWARD

The idea for this book actually came to me in the winter of 2006 when my wife, Karen, gave me a video camera and suggested that I go over to my dad's house to interview him about the colorful life he was living. However, a few days later before I could follow through, Dad passed away at the age of 88. Years later, after the strong urging of my family, I sat down to write this book with only memories, editing help from my wife Karen (she softened and enhanced it a bit) and some research to guide me. Most of the book is based on actual events and our experiences on the whaler *Dennis Gayle*. Except for my brother, Bill, there is no one to say things didn't happen exactly as I described. I've changed the last names of some of the characters in the book to protect the privacy of their families.

After his whaling career, Gib Hunter went on to forge new grounds in the commercial fishing industry. He owned nine fishing boats which delivered seafood product to the Eureka Fisheries processing plant in Fields Landing. The processing side of the business, led by Gib and his General Manager, Budd Thomas, employed hundreds of people in seven locations along the West Coast and processed 25 million pounds of fish a year.

On May 12, 1994, Gib donated his former research vessel, the *Pacific Raider* to Humboldt State University. According to the University President Dr. Alistair W. McCrone, "the gift of the 135-foot *Pacific Raider* to Humboldt State University elevates the instructional and research capabilities of Humboldt's marine science program. The *Pacific Raider*, now re-commissioned as the *Pacific Hunter* affords Humboldt State one of the largest university-owned academic vessels in the nation."

In 1995, Gib received a national award as he was named Person of the Year by the National Fisheries Institute (NFI). The NFI represents 1,000 companies involved in the fishing industry, from harvesting companies to grocery stores. He was honored for his many years in the industry, as well as his work studying fish stocks off the coast of the Pacific Northwest.

At the Boston, Massachusetts ceremony, the NFI President said that Gib Hunter also worked at developing markets for fish species that didn't have a strong market. He continued that "Gib is one of the last pioneers of the fishing industry. He's not only trying to make a living for himself, but he's trying to help the industry grow and diversify."

The last few years of his life, Gib could be found looking over his son Bill's shoulder as Bill and his son Travis managed his fishing fleet in Fields Landing. Or, Gib would be skippering his yacht, the *Gibbershea* off the Florida coast. His home was the sea. He would be most comfortable puffing on a cigar in the wheel house or the bridge of his yacht as it slowly cruised toward the Bahamas and beyond.

Two days after Gib passed away, an unpredicted storm hit Humboldt Bay on the north coast of California. The boat he had since he was eighteen, the *Winga*, after a severe battering, sank at the dock. Was it a coincidence? We'll never know.

ABOUT THE AUTHOR

After braving the open ocean during the whaling years, Dennis Hunter moved in a different direction. He attended Humboldt State University with a major in Radio and Television and spent a few decades in a broadcasting and marketing career. He currently lives in Eureka, California with his lovely wife, Karen and their two four-legged kids. They have four children, eight grandchildren and six great-grandchildren. After writing 30-second and 60-second commercials, this is his first and only venture in a new venue.